Apple Seeds

Katie Peters

GRL Consultants,
Diane Craig and Monica Marx,
Certified Literacy Specialists

Lerner Publications ◆ Minneapolis

Note from a GRL Consultant
This Pull Ahead leveled book has been carefully designed for beginning readers.
A team of guided reading literacy experts has reviewed and leveled the book to
ensure readers pull ahead and experience success.

Lerner Publications Company
A division of Lerner Publishing Group, Inc.
241 First Avenue North
Minneapolis, MN 55401 USA

For reading levels and more information, look up this title at www.lernerbooks.com.

Main body text set in Memphis Pro 24/39
Typeface provided by Linotype.

Photo Acknowledgments
The images in this book are used with the permission of: © Shutterstock (all)

Front cover: © Shutterstock

Library of Congress Cataloging-in-Publication Data

Names: Peters, Katie, author.
Title: Apple seeds / Katie Peters.
Description: Minneapolis : Lerner Publications, [2020] | Series: Science all around
    me (Pull ahead readers - Nonfiction) | Includes index. | Audience: Age 4–7. |
    Audience: K to Grade 3.
Identifiers: LCCN 2018058176 (print) | LCCN 2019000017 (ebook) |
    ISBN 9781541562264 (eb pdf) | ISBN 9781541558496 (lb : alk. paper) |
    ISBN 9781541573314 (pb : alk. paper)
Subjects: LCSH: Apples—Development—Juvenile literature.
Classification: LCC SB363 (ebook) | LCC SB363 .P386 2020 (print) | DDC 634/.11—dc23

LC record available at https://lccn.loc.gov/2018058176

Manufactured in the United States of America
1 – CG – 7/15/19

Contents

Apple Seeds 4

Did You See It? 16

Index 16

Apple Seeds

This is an apple seed.

This is an apple tree.

This is an apple flower.

This is a little unripe apple.

This is a big ripe apple.

The apple has seeds
inside it.

Did You See It?

apple

flower

seed

tree

Index

apple flower, 9

apple seeds, 5, 15

apple tree, 7

ripe apple, 13

unripe apple, 11